Cement Types, Admixtures, and Technical Procedures of Cement Analysis

An Introduction

Synthesis Lectures on Chemical Engineering and Biochemical Engineering

Editor
Robert Beitle, Jr., ***University of Arkansas***

Cement Types, Admixtures, and Technical Procedures of Cement Analysis: An Introduction
Tadele Assefa Aragaw
2020

Informing Chemical Engineering Decisions with Data, Research, and Government Resources
Patricia Elaine Kirkwood and Necia T. Parker-Gibson
2018

Cement Types, Admixtures, and Technical Procedures of Cement Analysis: An Introduction

Tadele Assefa Aragaw

www.morganclaypool.com

ISBN: 9781681736532 paperback
ISBN: 9781681736549 ebook
ISBN: 9781681736556 hardcover

DOI 10.2200/S00947ED1V01Y201908CHE002

A Publication in the Morgan & Claypool Publishers series
SYNTHESIS LECTURES ON CHEMICAL ENGINEERING AND BIOCHEMICAL ENGINEERING

Series Editor: Robert Beitle, Jr., *University of Arkansas*
Series ISSN
Print 2327-6738 Electronic 2327-6746

Cement Types, Admixtures, and Technical Procedures of Cement Analysis

An Introduction

Tadele Assefa Aragaw
Bahir Dar Institute of Technology, Bahir Dar University, Bahir Dar, Ethiopia

SYNTHESIS LECTURES ON CHEMICAL ENGINEERING AND BIOCHEMICAL ENGINEERING #2

ABSTRACT

A bulky document on cement science and manufacturing technology is difficult for a college junior to easily understand. Thus, it is better to write a short and precise book that contains only the necessary basic content.

This introductory book is designed as a short and concise resource for undergraduate university students studying chemical science (chemistry and chemical engineering), material science, geology, and construction technology. It emphasizes different types of cement, admixtures, and how to analyze the chemical compositions of cement in the laboratory. Technical procedures of cement analysis are very important for determining and comparing chemical compositions. This book describes the detailed procedures for different test parameters.

KEYWORDS

types of cement, cement admixture, cement test, methods of analysis, cement characteristics

Contents

Preface

This book is a follow-up to my previous book, *Concise Introduction to Cement Chemistry and Manufacturing*. Here, however, more emphasis is placed on the chemical analysis of cement vs. the underlying chemistry and manufacturing process of cement. In order to help the readers understand the context in which this book has been bulleted for chemical engineering, civil engineering, industrial chemistry, and more on the cement analyst, the book represents a summary of information collected from the limited number of sources. It was written based upon the author's understanding of the introductory sciences surrounding cement and its chemical analysis, specific tests, and methods. The information provided in this book is intended as a resource to determine the principles of chemical analysis parameters as well as the detailed laboratory procedures for cement testing. Chapter 1 presents general information about raw materials for cement making, naturally occurring raw materials, industrial waste products, and the chemistry of cement production. Chapter 2 describes in more detail the different cement types and their major characteristics. Chapter 3 presents different types of admixtures with their characteristics as well as the crystalline and colloidal theory of cement. The last chapter, Chapter 4, which is the major section of this book, describes the chemical analysis of cement with detailed laboratory procedures. This book outlines the best available techniques and is up-to-date as of its publication, however it will be reviewed and updated as appropriate based on future scientific advancements.

Tadele Assefa Aragaw
November 2019

Acknowledgments

I would like to send my grateful appreciation to Mr. Fekadu Mazengiaw for his intensive review of all the chapters. As for the book itself, I received welcomed comments, encouragement, and/or help from the instructors and researchers of the Faculty of Chemical and Food Engineering, Bahir Dar Institute of Technology, Bahir Dar University. I hope that the remaining errors are minor ones and will not disturb the reader.

Tadele Assefa Aragaw
November 2019

CHAPTER 1
Introduction

Cement is a finely powdered substance which possesses strong adhesive powers when combined with water, creating a chemical reaction.

The word "cement" comes from the Latin word "cementum" which means pieces of rough, uncut stones and the Ethiopian word "ciminto" which might have originated from "cemento" (chemento), the Italian word for cement.

The Romans made cement that would harden under water by mixing lime and ground volcanic rock, crushed pottery, or rubble. The first great improvement in cement making is attributed to civil engineer John Smeaton (1724–1792), who undertook the construction of a new lighthouse on eddy stone rocks in the United Kingdom in 1756. He found that the best mortar for work underwater could be made by burring limestone that contained a considerable amount of clay material. In taking the stone from the quarry, Smeaton chose those layers that had been shown by testing whether it can be burn to lime or not can be dispersible with water. He had grounded the unshakable portions by producing a cementing material having much more hydraulic properties, because of this product can have higher surface area which can easily hydrated with water molecules.

About 50 years after Smeaton's discovery, Louis Vicat, a French chemist, found that it was not necessary to depend upon the rock in which limestone and clay occurred coincidentally. Vicat produced a hydraulic cement by burning finely ground chalk and clay that had been mixed and made into a paste. The first patent granted for the making of Portland Cement was issued in 1824 to Joseph Aspdin, an Yorkshire brick layer, who heated finely powdered chalk, or road dust from a limestone road, with the clayey mud of the river Medway. Because the product bore a slight resemblance to a well-known limestone from Portland, England, called Portland Stone, he named his product "Portland Cement." It was first manufactured in the United States in 1875 at Coplay, and it was the first manufacturing product next to iron in the world.

Most of the time when people mention cement they are referring to Portland cement, which has become part of the international nomenclature. Although this particular type of building material is governed by standard specifications which vary from country to country, its general proprieties are more or less the same all over.

Cement manufacturers have a rather wide choice of raw materials. Lime, silica, and alumina are the most important ingredients and any materials that supply these compounds can be made into cement manufacture provided that they do not contain excessive amounts of other oxides.

Based on the kind and amount of composed substances, different types of cement are known. As mentioned previously, the most commonly known cement throughout the world is called Portland cement which contains lime, silica, and alumina which, when burned together, form tricaleic silicate, 3 CaO, SiO_2, tricalcium aluminate, 3 CaO, Al_2O_2, and dicalcium silicate, 2 CaO silicate, 2 CaO, SiO_2; these are unstable compounds, which once wet react at different speeds.

A cement is analyzed to keep the qualities, that is the right proportion of the constituents, since the properties of cement are determined by its constituents, for example, setting time, amount of heat of hydration, resistant to the attack of chemicals especially of sulphates and of seawater, the color of cement produced, soundness, etc., are all dependent on the composition of the cement. As described earlier, the principal compounds present in Portland cement are tricalcium silicate (C_3S), dicalcium silicate (C_2S), tricalcium aluminate (C_3A), and calcium alumina ferrety (C_4AF). However, the compositions of cement are calculated from the analytical results expressed in terms of percentages of calcium oxide (CaO), silica (SiO_2), alumina (Al_2O_3), and iron oxide (Fe_2O_3).

But there are problems in analyzing these compositions, i.e., in determining the exact composition of the cement produced, for example, variations in chemical composition of raw materials happens from time to time, interferences occur during any series of analysis (e.g., analysis on the raw materials and analysis on the final product, i.e., the cement), etc.

1.1 RAW MATERIALS FOR CEMENT MAKING

1.1.1 NATURALLY OCCURRING RAW MATERIALS

The exploration and exploitation of better raw materials were necessary for the development of cement manufacturing.

The main component of most cement types are lime, silica, and alumina. When combined, those elements form tricalcium silicate $3\,CaO \cdot SiO_2$, tricalcium aluminates $3\,CaO \cdot Al_2O_3$, and dicalcium silicate, $2\,CaO \cdot SiO_2$.

Portland cement is by far the most abundantly used cement worldwide. The chemical composition of Portland cement clinker indicates that the raw material component should be predominantly calcareous (lime suppliers), with successively smaller amounts of silica, alumina, and iron-rich constituents. The occurrence of rocks which on burning without any admixture that give cement clinker is extremely rare. Among the very few occurrences, mention may be made of the "Portland stone" in Lehigh, Ohio (USA) and Novorossiysk (Russia). The choice of raw materials, therefore, falls on verity of naturally occurring rocks or industrial waste. For major calcareous components, the choice is preferably a naturally occurring, well-explored limestone deposit because of its easy and alluring availability in most cases. Generally, to produce 1 ton of cement, about 1.6 tons of raw materials are needed, out of which the calcium carbonate component may constitute 70–99.5%, depending upon quality. Besides limestone, industrial wastes are also considered as raw materials, when availability is easy, and these wastes fit in well with other economic and technical parameters when considering the planning of a cement plant.

When considering the planning of a cement plant, it is rather difficult to get a source of raw materials which have an approximate composition of the proper raw materials, even though this has an advantage of lesser involvement with chemical substances, which are usually added to a clinker. A better choice is to start with a pure form of limestone and then mix it with necessary additives. Experience shows that, even with the most modern and sophisticated techniques of geological prospecting and evaluation, the quality characteristics and their variation within a buried natural mineral deposit can be at best relied upon but not fully guaranteed.

Limestone and clay are the required materials for cement. The limestone chosen should be such that it contains the necessary constituents in proportion so near to the actual chemical requirement that only slight adjustments are necessary. Therefore, the manufacturing plant should be constructed as near as possible to the deposit.

The materials are blasted in the quarry and brought to a gyratory crusher that reduces the rock to a size of an egg. The addition of a correct amount of limestone takes place at this stage. The relative proportions of raw materials include: lime (CaO), silica (SiO_2), alumina (Al_2O_3), magnesia (MgO), iron (Fe_2O_3), sulphur trioxide (SO_3), and alkali (Na_2O, K_2O).

1.1.2 INDUSTRIAL WASTE PRODUCTS

When manufacturing cement there should be a proportional amount of raw materials. However, during the feeding of the kiln by raw materials some are lost due to the volatility of the substance. For example, during calcination, the thermal dissociation process involving the separation of volatile from nonvolatile components in the furnace, substances which are volatile are lost, so that there is a shrinkage in volume in the material obtained. Therefore, these lost materials can be adjusted by some industrial waste products. Slag, clay, and limestone are some of the industrial wastes that can be used to adjust types of cement.

Slag

Slag from an iron blast furnace may be adjusted to suit the manufacturing of Portland cement as it is required for specific purposes. It is mixed with limestone to make the desired percentage. The shrinkage, in this case, is also somewhat less. The raw materials may be purified by the fourth floatation process. That is the process by which low-grade ores are separated from gangue as a result of preferential wetting action.

Slag is the impure material removed in the process of making pig iron, and in smelting copper, lead, and other materials. The slag from steel blast furnaces is called clinker. It contains silicates of calcium, magnesium, and aluminum. The slag from a copper and lead smelting furnace contains iron silicate and other metals in small amounts. Slag from an open health steel furnace contains lime and some iron often smelted again. As described previously, since the slag contains the essential ingredients which are also in cement, it is, therefore, used to create a cement of the desired percentage.

Clay

In general, the term clay implies a natural earthy fine-grained material which develops plasticity when mixed with a limited amount of water. Plasticity refers to the property of moistened material to be deformed under the application of pressure, and the deformed shape is retained when the deforming pressure is removed.

Chemical analysis of clay shows it to be essentially consisted of silica, alumina, and water frequently with appreciable quantities of iron, alkalis, and alkaline earth. Quantitative chemical analysis alone does not explain the differences in properties that some of the clay exhibits, although their composition, as found by analysis, is the same. X-ray diffraction (XRD) techniques with an electron microscope demonstrate that clays have a crystalline structure. The great variation in properties can be accounted only by assuming that the clay minerals are composed of comparatively simple building units and that the difference is primarily due to the way these units are put together.

The properties of clay materials are controlled by at least five major factors, and these attributes which characterize clay materials are:

- clay-mineral-composition,
- non-clay-mineral-composition,
- organic materials,
- soluble salts, and
- exchangeable ions.

Generally, the clay mineral compositions are the most important factors and sometimes as little as 5% of a particular clay mineral may largely determine the properties of the whole clay.

When clay materials are heated, dehydration occurs. Thus, there is a loss of any water (adsorbed, interlayer, or structural HO^- water heated by the clay minerals). The heating of clay materials also causes changes in the clay materials structures.

The economic use of clay material is determined largely by its clay mineral composition. In construction engineering, a knowledge of the clay material on or through which a structure is to be built is essential. Commercial clay or clay utilized as a raw material in manufacturing is among the most important non-metallic mineral resource.

Limestone

Limestone is a sedimentary rock composed dominantly of carbonate minerals, principally carbonate of calcium and magnesium. Limestone is the most abundant non-elastic rock and is overwhelmingly the largest reservoir of the element carbon at or near the surface of the earth.

Although the word limestone is used in a general sense, it is specially referred to as carbonates rock eliminated by the mineral dolomite, $CaMg(CO_3)_2$. Although the mineralogical composition of most limestone is similar, its texture, a term denoting primarily absolute and relative to size and shape of the visible constituents of the rock, is not similar because it is formed under a great variety of conditions.

The chemical composition of limestone is largely calcium oxide, carbon dioxide, and magnesium oxide; if magnesium oxide exceeds 1 or 2%, the rock may be referred to as magnesium limestone ($MgO \cdot CaCO_3$). A small amount of silica and alumina may be also present as a result of the presence of elastic minerals, quartz, and clay. The iron oxide may be present either as carbonate (siderite, $FeCO_3$), ferroan-dolomite, $Fe_2O_3 \cdot MgCa\,(CO_3)_2$, or in other minerals such as clays. Strontium may be present as an important trace element, probably derived from the original fossil material in which it was incorporated into aragonite, $CaCO_3$.

The chief mineral of limestone is calcite and aragonite, and in the dolomite limestone, dolomite calcite, and aragonite have the same composition but a different crystal structure. Aragonite is unstable with reference to calcite in the surface environment, and it is transformed into the calcite with time. Even though aragonite is unstable, it forms as precipitates from sea water and some fresh water. Although dolomite is stable in a surface environment, it is primarily precipitated only under special conditions, such as high salinity or high alkalinity.

Siderite, the iron carbonate, is found in some limestone, but iron occurs in carbonate form mainly as ferro-dolomite. Silica may be present either as finely disseminated throughout the rock or as segregated into modules chart. Silica also occurs as small crystal quartz that has grown in place during digenesis (the gradual and successive chemical and physical changes which take place in sediments previous or during their consolidation). Feldspars occur in the same way but are a little less common than silica. Other minerals found in limestone are glauconitic, hydrous-silicate of potassium, and iron of somewhat variable composition, collophane, $Ca_{10}(PO_4Ca_3)_6F$, and pyrite, FeS_2. A host of other minerals may be found as small amounts of elastic materials brought in by currents including almost always as small amounts of fine grained clay.

1.1.3 THE CHEMISTRY OF CEMENT PRODUCTION

Portland cement is the most commonly used cement for construction. However, there are other types of cement available for other purposes. Cement is very versatile due to its property, strength, and simplicity, and has, therefore, revolutionized the concept of construction.

The main constituents of cement are tricalcium silicate ($3\,CaO \cdot Al_2O_3$) and tetracalcium alumino ferrite ($4\,CaO \cdot Al_2O_3Fe_2O_3$) which is formed at different burning temperatures. For example:

- below 800°C, the formation of $CaO \cdot Al_2O_3$ and probably $CaO \cdot Fe_2O_3$;
- 800–950°C, the formation of $5\,CaO \cdot SiO_2$ and $CaO \cdot SiO_2$;
- 950–1200°C, the formation of $2\,CaO \cdot SiO_2$;
- 1200–1300°C, the formation of $3\,CaO \cdot Al_2O_3$ and probably $4\,CaO \cdot Al_2O_3 \cdot Fe_2O_3$; and
- 1260–1450°C, the formation of $3\,CaO \cdot SiO_2$ with disappearance of free CaO.

The chemical compounds that have been identified in Portland cement clinker, together with the names by which some of them were formerly known, as well as the abbreviations by which they are usually designated, are shown in Table 1.1.

Tricalcium silicate acts rapidly, forming gelatinous calcium hydrate and gelatinous silica, after having been wetted. The hydration continues, the gelatinous material binds the grains of sand which are always added, and the crushed stone filler, to a hard mass. Tricalcium aluminate acts faster than tricalcium silicate, but does not produce a strong bond. Dicalcium silicate acts only after months have passed. All three compounds release heat when mixed with water. The maximum heat release compounds are the aluminates that are responsible for most of the undesirable properties. Cement having less aluminate has less

Table 1.1: Chemical compounds of Portland cement and their names

Formula	Abbreviations	Name
$3CaO.SiO_2$	C_3S	Alite
α $2CaO. SiO_2$	C_2S	Belite
β $2CaO. SiO_2$	C_2S	Felite
$4CaO.Al_2O_3.Fe_2O_3$	C_4AF	Celite
$3CaO.Al_2O_3$	C_3A	
$5CaO.SiO_2$	C_5S	

initial strength but higher ultimate strength. Also, there will be less generation of heat, more volumetric stability, less cracking, and more resistance to acid attacks.

The scheme and equipment used for making Portland cement is representative of the majority types of cement. Portland cement is made by firing a careful proportional mixture of clay and calcium carbonate to the point of sintering.

The following reaction gives some idea of the chemical change that takes place during the manufacturing of cement. Formation of cement takes place by the reaction of the solid-state to a greater extent. Firing reaction:

$$\mathrm{CaCO_3} \quad \longrightarrow \quad \mathrm{CaO + CO_2}. \tag{1.1}$$

$$\text{Clay} \quad \xrightarrow[-\mathrm{H_2O}]{500°\mathrm{C}} \quad \text{dehydrated Clay} \quad \xrightarrow[\text{reactants}]{650°\mathrm{C}}, \quad \text{a mixture of } \mathrm{Al_2O_3} \text{ and } \mathrm{SiO_2}. \tag{1.2}$$

Then,

$$\mathrm{CaO + Al_2O_3} \quad \longrightarrow \quad \mathrm{CaO \cdot Al_2O_3}. \tag{1.3}$$

$$\mathrm{CaO + SiO_2} \quad \longrightarrow \quad \mathrm{2\,CaO \cdot SiO_2}. \tag{1.4}$$

$$\left(\text{Or dehydrated clay} + \mathrm{CaO} \quad \xrightarrow{650°\mathrm{C}} \quad \mathrm{CaO \cdot Al_2O_3 + 2\,CaO \cdot SiO_2}.\right) \tag{1.5}$$

Finally,

$$2\,CaO + CaO \cdot Al_2O_3 \longrightarrow 3\,CaO \cdot Al_2O_3. \tag{1.6}$$

$$CaO + 2\,CaO \cdot SiO_2 \longrightarrow 3\,CaO \cdot SiO_2. \tag{1.7}$$

The hydraulic reactivity of Portland cement is especially dependent on crystal structure and the defects/disorder in crystals.

Since Portland cement is a polyphase inorganic binder, i.e., the principal phases present in it are: alite (C_3S), belite (C_2S), aluminate (C_3A), and alumino-ferrite (C_4AF). The hydraulic properties of cement are to a large extent dependent on these phases. For example, the crystal structure of C_3S is trigonal. There are three holes in the structure per formula weight. But tricalcium aluminate, a definite compound whose crystal lattice is cubic, have eight holes per formula unit in the structure. Therefore, the presence of these holes increases the absorption of water molecules, hence greater hydraulicity.

High hydraulicity of cement phase, for example, C_3S, B-C_2S, even C_3A, is also generally being credited to the defects in the crystal structure. These defects may arise due to substitution of foreign ions in crystal lattice. The high reactivity of C_3S has been associated with holes in the structure. Similarly, the high reactivity of C_3A can be assigned to holes in the structure. However, there are some foreign ions that may lower the hydraulic activity, for example, the introduction of Na in the C_3A structure leads to a more compact structure resulting in low hydraulic activity. The holes may be filled in by Na in C_3A structure which leads to a more compact structure resulting in low hydraulic activity. The order of hydraulic reactivity of cement compound in a pure state can, therefore, be arranged in the following way:

$$C_3A > C_3S > C_4AF > B - C_2S. \tag{1.8}$$

All these compounds are hydraulic and contribute to the strength of cement. The silicate phases, however, account for the major cementations materials with a moderate development of heat hydration, as shown in Table 1.2. In cement, there are not only the compounds indicated in Table 1.2 but also others like MgO, alkalioxide, and SO_3.

Magnesium oxide or magnesia (MgO) in Portland cement is derived from magnesium carbonate present in the original raw meal ingredients used in the manufacture of Portland cement clinker. Since limestone constitutes about 77%

Table 1.2: Heat of hydration

	J/g
C_3S	500
B-C_2S	250
C_3A	1340
C_4AF	420
Free CaO	1150
Free MgO	840
Ordinary Portland cements	375–525
Sulphate-resisting cements	355–440
Pozzolanic cement	315–420
High-alumina cement	545–585

of the raw meal by weight, the major contribution of MgO comes from magnesium carbonate present in it. At the clinkering temperature of about 1,450°C at which the raw meal is generally burned, the hydration of MgO in cement to $Mg(OH)_2$ causes a volume expansion. For this reason, limits on MgO content in Portland cement are given in most national manufacturing standards.

Alkalis and sulphates enter the clink as secondary components from the raw meal as shown in Table 1.3.

Table 1.3: Secondary components introduced by the raw meal

	g/kg of clinker
K_2O	7.8–31.2
Na_2O	0.8–9.4
SO_2	0.5–11.0

These constituents occur in the vapor phase as well as in the condensed phases. Because of this dual role, depending on the characteristics of the source and burning conditions, these constituents affect kiln operation and product quality.

CHAPTER 2

Types of Cement

Depending on their uses and properties, types of cement are divided into three main groups:

- air cements,
- acid-resistant cements, and
- hydraulic cements.

2.1 AIR CEMENTS

Air cements are one of the types of cement which can only harden and retain their strength in the air. These kinds of cements include pure and high calcium of white lime which has high calcium oxide content and is dependent on setting and hardening solely on the absorption of carbon dioxide from the atmosphere. They are used for making buildings, plastering mixes (lime), structural and decorative articles.

2.2 ACID-RESISTANT CEMENTS

Acid-resistant cements are types of cement, which after hardening withstand the action of acid materials. Acid resistance cement is made without firing from liquid, or soluble, glass (an aqueous solution of the silicates of alkali metals with a common formula $(K, Na)_2O \cdot NSiO_2$), finally ground acid resistant aggregates.

Cement powder consists of a mixture of pulverized aggregate and sodium fluosilicate. When this mixture is combined with liquid glass, the mass formed soon sets and then rapidly hardens. Setting and hardening take place as a result of the reaction between the liquid glass, and sodium fluosilicate leading to formation of a silicon acid gel $[H_4SiO_4]$ which possess bonding properties.

Acid-resistant cements are used for lining chemical equipment and for preparing mortar and concretes.

2.3 HYDRAULIC CEMENTS

Hydraulic cements are types of cement that can harden and retain their strength in water. They react exothermically with water to harden strong masses having extremely low solubility. A number of calcium silicate, calcium aluminates, and related compounds are able to react in these way and they are the active components of commercial hydraulic cement. The world demand for this cement is measured in hundreds of millions of tons per year, which means that they have to be produced from naturally occurring raw materials rather than from pure chemicals. Because impurities are associated with the mineral deposits used, commercial hydraulic cements contain a range of active compounds rather than one compound alone. Modern production techniques make it possible to favor the formation of the desired compound and suppress that of an undesired one, thus to control the composition and properties of cements.

2.3.1 ROMAN CEMENTS

Concrete has been used as a construction material for centuries. Before 100 B.C. the Romans had developed an excellent concrete which enabled them to erect vast structures and works of engineering.

On the slope of Mt. Vesuvius and, in extinct volcanic areas near Rome, they found a light porous volcanic rock. Its rough surface formed a good bond for cementation materials (a substance capable of acting as cement) or mortar. The cement was prepared from a mixture of lime and a volcanic ash called pozzolana, named after the village of Pozzuolic near Mt. Vesuvius. When the pozzolana was mixed with limestone and burned, the resulting materials, ground and mixed with water, gave the cement an unprecedented strength.

2.3.2 NATURAL CEMENTS

Certain natural rocks when quarried (taken stones of various size from natural rocks), crushed, and processed, will produce natural cement. If enough heat is applied to drive off gases, a hydraulic cement result, but it has very low strength. Lime and natural rock were the only sources of cementation materials for many centuries. Because the strength and other physical properties of natural cements vary greatly, very few building codes allow its use in concrete.

2.3.3 PORTLAND CEMENTS

Portland cement is a greenish grey impalpable powder. Its main constitutes are lime, silica, alumina, and iron oxide. It was first manufactured in the United States, in Pennsylvania in 1872. It was discovered that if a carefully controlled mixture of limestone and clay was burned at a much higher temperature than it was ever used before, the resulting cement had better hydraulic qualities. Most of the ingredients in Portland cement are found in nature, however, they cannot always be used in natural form. It is suitable for the most types of work and thus, it is the most widely used of the hydraulic cements.

Over the years, several varieties of Portland cement which exhibit special characteristics or properties, that are of value in appropriate circumstances, have been developed. They have in common the fact that they all contain the same active minerals only the proportion of each is different.

Rapid Hardening Portland Cement

This cement gains strength faster than ordinary Portland cement. It is similar in composition to ordinary Portland cement but it is more finely grounded. This does not render it quick-setting, and concretes made with it remain workable for periods similar to those of concrete made with ordinary Portland cement. The finer grinding does, however, increase the rate of hydration at an early age, and this leads to the increase in rate of hydration and early hardening implied by the name. This property is particularly useful when concrete is required to develop enhanced strength ages, for example when it is necessary to minimize the period of which formwork must remain in position. The increased rate of heat evolution associated with this cement may also be useful when concreting in water.

Sulphate-Resisting Portland Cements

Sulphate-resisting Portland cement (SRPC) is a form of Portland cement with low tricalcium aluminates (C_3A) content. It usually has a higher content of tetracalcium aluminoferrite (C_4AF) than other Portland cements, which gives it a darker color. Because sulphates can react with hydrated tricalcium aluminate, causing weakening, concrete made with this cement is more resistant from attack by the sulphate component which may be found dissolved in groundwater and also present in sea water.

Concrete made with SRPC has been found to be satisfactory in nearly all troublesome conditions when used below ground concreting. However, resistance to sulphate attack depends on the cement content and imperviousness of the concrete as well as on the concentration of sulphate encountered.

SRPC is similar to other Portland cement in the fact that is not resistant to acid, nor is it immune to the effects of some other dissolved salts, such as magnesium compounds which may occur in natural waters or effluents.

White Portland Cements

White cement is distinguished by its low content of iron compound which gives the grey-green color of ordinary cements. White cement is manufactured from white chalk and clay free from iron oxide. Gypsum is added to control setting, and special care is taken at all stages of processing not to introduce colored contaminations. White cement is used for decorative purposes and also for some colored concretes in which case a pigment is added to the mix.

Low-Heat Portland Cement

The heat generated by white cement setting may cause the structure to crack in the case of concrete. Heat generation is controlled by keeping the percentage of tricalcium aluminate and of tricalcium silicate low, that is, more dicalcium silicate than ordinary Portland cement. Thus, it hydrates slower and releases heat less rapidly than the latter, that is, its initial and final setting times are nearly the same as those of ordinary cement but the rate of its developing strength is very slow.

It is intended primarily for use in massive constructions, where the rate of heating evolution associated with ordinary Portland cement at an early age might cause undesirable thermal stress in the immature concrete.

Air-Entrained Portland Cement

This cement is ordinary Portland cement mixed with small quantities of air entraining materials at the time of grinding. Usually, air entraining materials are oils, fats, and fatty acids. These materials have the property of air entraining in the form of fine air bubbles in concrete. These bubbles render the concrete more plasticity, making it more workable and more resistant to freezing. However, because of air entraining, the strength of concrete reduces, therefore, the quantity of air entraining should not exceed 5%. In this case, less water is needed to produce a workable concrete mix. By incorporating other materials during manufacturing, generally, when the clinker is being grounded, an even wider range of cement is produced.

Portland-Blast Furnace Cement

Portland-blast furnace cement is obtained by mixing Portland cement clinker, gypsum, and granulated slag in proper proportions and finely grinding it. This cement has very similar properties of those of ordinary Portland cement with the following improvements.

1. It has lesser heat of hydration.
2. It has better resistance to soils, sulfates of alkali metals, alumina, and iron.
3. It has better resistance to acidic waters.

Hydrophobic-Portland Cement

In this cement (available only to special order), the particles are coated during the manufacturing with a waterproof skin of oleic or other water repellent to resist hydration in case the cement has to be stored for a long time in a moist atmosphere. This coating is rubbed off by friction in the mixer, and the cement then behaves normally. In most cases, less than 0.5% of an additive is required to obtain the desired effect. The added material may also serve as plasticizer in the concrete mix.

Water-Repellent Cement

If a metallic soap is mixed to cement, concrete made with the cement becomes water repellent and tends to shed and drain rain water better than normal concrete. This is particularly important with decorative concrete and case stone which has a pore structure. It is available and is used in precast concrete as well as in other branches of construction industry. Water-repellent property can be combined in the same cement.

Alumina Cements

Alumina cement is made by firing a mixture of bauxites and limestone to the sintering point or, more often, to the melting point, followed by fine-grinding produced clinker. This cement contains 35–55% of Al_2O_3, mainly in the form of CaO Al_2O_3, 35–45% of CaO, 5–10% of SiO_2, and up to 15% of Fe_2O_3.

Alumina cement is an extremely rapid setting cement material. After three days of hardening its strength is equivalent to that of a 28-day-old Portland cement. Alumina cement is, therefore, used in those cases where a structure must be urgently put into operation, for instance, in express type construction, or in emergencies. A large amount of heat is liberated when this cement harden, which is especially valuable when carrying out building operations in the winter. However, the cast of alumina cement is substantially higher than that of Portland cement.

Quick-Setting Cement

Quick-setting cement sets faster than the ordinary Portland cement. Its initial setting time is 5 min and final setting time is 30 min. It is used for making concrete that is required to set early, as for lying underwater or in running water. Initial setting time is very short, there is always the danger that the concrete will undergo initial setting during mixing and placing.

Colored Cement

By mixing suitable pigment, ordinary Portland cement could be given a red or brown color. For other colors, 5–10% of the desired pigments are grounded with white cement. Pigments used in cement should be chemically inert and also durable so as not to fade due to the effect of light, sun, or weather.

Portland Pozzolana Cement

This cement has properties similar to those of ordinary Portland cement, and can, therefore, be used for all general purpose where the latter is employed, or intimately and uniformly blending Portland cement and fine pozzolana. This material is of volcanic origin containing various fragments of pumice, pyroxenes, quartz, etc.

- Pumice: when the mortar rock or lava is impregnated with steam bubbles, it forms glass of cellular characteristics known as pumice. They are silicate.

- Pyroxenes: it is a very common group of minerals of many varieties, silicates of magnesium, iron, calcium, and other elements as important constituents of many kinds of rocks. These materials contain compounds of silicon, aluminum, and some other elements.

Definition—Isaac Lea (1792–1886), American publisher, defines pozzolana as siliceous materials, which are not cements in and of themselves, contain components, which in ordinary temperatures will fuse either processed or unprocessed and when finely grounded with lime in the presence of water will form compounds which have a low solubility and contain cementing properties.

The main use in replacing a portion of cement when making concrete and the principal advantages to be gained are economic, improvement of the workability of the concrete mixture, and reduction of bleeding and segregation. Being faster, they can be easier shaped. Other advantages are greater imperviousness, resistance to freezing and thawing, and resistance from attack by sulphates and natural waters. They reduce the disruptive effects of alkali aggregate reactions and also that of heat hydration.

When pozzolanas are mixed with cement hydration reaction is more rapidly than without pozzolanas. The amorphous forms of silica pozzolanas reacted with lime more rapidly than the crystalline forms of silica's. This is because of the pozzolanas chemical activity.

It is commonly thought that the lime silica reaction is the main or only one that takes place, but recent information indicates that alumina and iron, if present, also take part in the chemical reactions. It was observed that when pozzolana containing high percentage of iron and aluminium was used it can cause a high reduction in alkalinity with only a small silica release. This may be due to the fact that alumina and perhaps iron may be more important than silica in the reactions.

CHAPTER 3

Admixture and Characteristics of Cement

Admixtures are a wide range of chemical substances which are added to various types of end products, usually in rather low percentages, to provide some unique properties or to stabilize them against spoilage or deterioration.

Various admixtures (additives) are introduced to the types of cement that give them the required properties and also to reduce the manufacturing costs. If types of cement have additives, it delays the chemical reaction that takes place when the concrete starts the setting process, it improves workability, etc.

3.1 DIFFERENT TYPES OF ADMIXTURES

3.1.1 AIR-ENTRAINED ADMIXTURE

These can be obtained by the addition of an air-entraining agent during the manufacturing of cement. In this case the admixture used are oils, fatty acids, etc., this admixture can be incorporated in three main ways:

1. the addition of chemicals such as aluminum powder or zinc powder;
2. the use of cement-dispersing agents which are surface active chemicals that cause electrostatic changes imparting to the cement particles, rendering them mutually repellent and thereby preventing coagulation; and
3. means of surface active agents which reduce surface tension.

3.1.2 RETARDERS

These normally increase the setting time and reduce the subsequent rate of hydration and consist of calcium, sodium, potassium, or aluminum salts of lingo sulphonic acid, hydroxyl-carboxylic acids, and their salts and carbohydrates. The mechanisms of the delay is set, that is, in the best-made cement exists a little free lime and also it is present a little alumina-free lime that dissolving rapidly, retards the hydration of the aluminates.

Gypsum ($CaSO_4 \cdot 2\,H_2O$) is also used as a retarder, a negative catalyst required for controlling the setting (the reactions of cement with water). The amount of gypsum required to produce the most favorable "time-of-set" (the time required for cement to react to water) has been found to depend on alumina. It is known that compounds of alumina, via tricalcium aluminates and tetracalcium alumino ferrite, react very quickly with water and are the first to hydrate. With the addition of gypsum, the alumina of the cement reacts with calcium sulphate to form calcium sulpho-aluminate which is insoluble and does not hydrate.

The particle size of gypsum is somewhat similar to that of the clinker. Ordinary 97.65 kg of clinker need 2.35 kg of gypsum, or for one barrel of cement, that is, 376 kg, the requirements are 376 kg of clinker and 9 kg of gypsum.

3.1.3 ACCELERATORS

Accelerators normally increase the setting and the subsequent hydration properties of the cement. The additive increases the rate of hydration and also increases the cement rate of gain of strength. They consist of alkali carbonates aluminates and silicate, aluminium chloriode, calcium chloride, sodium chloride, sodium sulphate, caustic soda, soda ash, and potassium sulphate. The best results are obtained by the addition of calcium chloride. It increases the workability of the mixture slightly so the water content can be reduced by up to 10%. In very cold weather, a very high percentage of calcium chloride is added with sodium chloride, of course, depending on the temperature. The action of calcium chloride plus sodium chloride is three fold.

1. They lower the temperature at which the freezing of the wet mixture takes place.

2. They help to keep the mixture warm by accelerating the generation of heat by chemical action.

3. They increase the ability to resist frets by speeding up the rate of strength. The accelerating effects of the additives (inorganic salts) are due to the formation of insoluble double salts, separated in the form of fine crystals.

By these crystal seeds the surface area is increased to a considerable extent and thus crystallization and concomitant hardening is hastened.

Cements with accelerator additives are useful in low temperature zones.

3.1.4 PIGMENTS

Pigments are used to color the types of cement. All pigments must be permanent. To produce an effective color they are generally grounded with the cements in the ball mill. They are used only in small quantities so they are mixed with fillers or extenders, for example, chalk and barium sulphate. They are usually metallic oxide of a non-folding nature. Cements made with such pigments are used for decoration purposes.

3.1.5 FINELY DIVIDED WORKABILITY ACIDS

These are all mineral powders which are grounded as fine as the cement or usually much finer. When concretes are made with such cement there will be an increase in workability by increasing the amount of paste and hence the cohesiveness. The materials used are lime, kaoline ($Al_2O_3 \cdot 2\,SiO_2 \cdot H_2O$), chalk, etc.

3.1.6 SURFACE ACTIVE AGENTS

Surface active agents are substances that increase the elasticity and bonding properties of the cement. They are used either as powder flakes or liquid and have been classified as:

1. synthetic detergents of alkyl-alkyl sulphonate type, the alkali group petroleum distillate such as kerosene while aryl group is usually a sulphonated benzene or naphthalene ring;

2. calcium-lingo-sulphonate derived from the sulphate process in papermaking, calcium chloride may also be added;

3. sodium salts of cyclo paraffine carboxylic acid; and

4. alkali or triethanolanine, salts of fatty acids derived from fats.

3.1.7 POZZOLANIC MATERIALS

Pozzolans are used as admixtures in place of a portion of the Portland cement. The pozzolanic materials most commonly used as admixtures are pumice (molten rock or lava) impregnated with steam bubbles; it is a silicate and not a silica volcanic ash, and calcined clays. Cements make the underground structure, but cannot be employed in conditions where temperature changes are in a wide range. Such cements harden very slowly. Certain aggregations, classified as reactive aggregates, react chemically with Portland cement and expands. A proven pozzolan may be used to control this expansion.

3.2 THE SETTING AND HARDENING OF CEMENT

The reaction between water and compounds present in a cement lead to the setting and hardening of the materials. The setting of cement is a chemical reaction called hydration; it evolves heat and is irreversible. No exact description of the process is possible, although much information has been gathered regarding the action of water on individual cement compounds. The effects of water are two kinds: (a) hydration and (b) hydrolysis. The hydration reaction is the one involved when on anhydrous salt such as copper sulphate reacts with water to form the familiar pent hydrate $CuSO_4$, H_2O. Hydrolysis is the term used to describe the reaction between water and salt which gives an acid and a base. Tricalcium aluminate reacts very rapidly with water giving a hydrated compound:

$$\mathrm{CaO \cdot Al_2O_3 + 6\,H_2O \quad \longrightarrow \quad 3\,CaO \cdot Al_2O_3 \cdot 6\,H_2O.} \tag{3.1}$$

This reaction is probably responsible for the preliminary set of the cement. Tricalcium silicate reacts slower and undergoes both hydration and hydrolysis.

Two theories have been advanced about the setting and hardening of cement.

3.2.1 THE CRYSTALLINE AND COLLOIDAL THEORY

According to this theory, the hardening of the cement is due to the interlocking of the crystal formation during hydration.

In addition, the hydration products of silicates from rigid cells and the development of strength is due to the hardening of the gel. This theory is satisfactory due to the following reasons.

- At ordinary temperature the hydrated calcium silicate appears to be non-crystalline.

- Many physical properties of set cement are explainable on the basis of this theory.

Therefore, the hydration of cement explains the hardening of cement.

In the hydration of cement by the addition of water into cement, the following two types of actions take place.

- Hydrolysis. When water is added to calcium, silicates hydrolysis takes place. Calcium silicates decompose into calcium silicate of lower basicity and release calcium oxide which forms calcium hydroxide with water.
  - Tricalcium silicate:

$$3\,CaO \cdot SiO_2 + H_2O \longrightarrow 2\,CaO \cdot SiO_2\,(ag) + Ca(OH)_2 \tag{3.2}$$

$$2\,(3\,CaO \cdot SiO_2) + 3\,H_2O \longrightarrow 3\,CaO \cdot 2\,SiO_2\,(ag) + 3\,Ca(OH)_2. \tag{3.3}$$

  - Dicalcium silicate:

$$2\,(CaO \cdot SiO_2) + H_2O \longrightarrow CaO \cdot 2\,SiO_2\,(ag) + Ca(OH)_2. \tag{3.4}$$

- Hydration. Silicates and aluminates take up water of hydration in the following manner.
  - Hydration of dicalcium silicate:

$$2\,CaO \cdot SiO_3 \longrightarrow 2\,CaO \cdot SiO_2 \cdot 4\,H_2O. \tag{3.5}$$

  - Hydration tricalcium aluminates:

$$3\,CaO \cdot AlO_3 \longrightarrow 3\,CaO \cdot Al_2O_3 \cdot 6\,H_2O. \tag{3.6}$$

  - Hydration of dicalcium aluminates:

$$2\,CaO \cdot Al_2O_3 \longrightarrow 2\,CaO \cdot Al_2O_3 \cdot nH_2O, \tag{3.7}$$

  where n varies from 5–9.

– Hydration of tetracalcium aluminates:

$$4\,CaO \cdot Al_2O_3 \quad \longrightarrow \quad 4\,CaO \cdot Al_2O_3 \cdot nH_2O, \tag{3.8}$$

where n varies from 12–14.

– Hydration of tetracalcium alumina-ferrite:

$$\begin{aligned} 4\,CaO \cdot Al_2O_3 \cdot Fe_2O_3 \quad \longrightarrow \quad & 3\,CaO \cdot Al_2O_3 \cdot 6\,H_2O \\ & + CaO \cdot Fe_2O_3\,(aq). \end{aligned} \tag{3.9}$$

On hydration of cement, there is an evolution of heat that occurs and, thus, there is a rise in temperature. This evolution of heat occurs during the first seven days of hardening. The following compounds are responsible for the evolution of heat in descending orders

1. tricalcium aluminates;
2. tricalcium silicate;
3. tetracalcium alumina ferrite; and
4. dicalcium silicate.

The heat evolved is used in the setting of the cement. The following are the functions of different compounds in setting of the cements:

- Tricalcium aluminates—is responsible for initial set.
- Tricalcium silicate—is responsible for first strength at 7 or 8 days.
- Dicalcium silicate and tricalcium silicate—are responsible for the final strength in one year.
- Fe_2O_3, Al_2O_3, Mg, and alkalis lower the clinkering temperature.

Initial Set and Hardening

A short time after the cement is mixed with water, hydration tricalcium aluminates are produced in amorphous form, and later crystallizes. At this time the low burned and finely grounded lime also is hydrated.

The next compound to react is the tricalcium silicate. Its hydration may begin within 24 hr, and is generally completed within 7 days. Between 7 and 28 days the amorphous aluminates begin to crystallize and dicalcium silicate begins to hydrate. Although the latter is the chief constituent of Portland cements (America), it is the least reactive compound.

The early strength (24 hr) of cement is probably due to the hydration of free lime and the aluminates. The increase of strength between 24 hr and 7 days depends on the hydration of aluminates that may contribute somewhat. The increase of strength between 7 and 24 days is due to the hydration of the calcium silicate, but here we encounter opposing forces in the hydration of any high-burned free lime present and in the crystallization of the aluminates. It is due to this hydration that the following off in strength between 7 and 28 days of very high-burned, high-limed cements is due, whereas the decrease shows by the high-alumina cements due to crystallization of the aluminates.

Retardation of Set by Gypsum

The exact manner in which gypsum retards the initial set of cement is not yet known. For a reason that is pointed out by Klein and Phillip's, the usual explanation that the retardation is due to the formation of the siphon aluminates crystal that has selective action on water is insufficient:

$$\begin{aligned} &3\,\mathrm{CaO}\cdot\mathrm{Al_2O_3}\,(\mathrm{ag}) + \mathrm{CaSO_4}\cdot 2\,\mathrm{H_2O} \\ &\quad\longrightarrow\quad 3\,\mathrm{CaO}\cdot\mathrm{Al_2O_3}\cdot\mathrm{CaSO_4}\cdot 3\,\mathrm{H_2O}. \end{aligned} \tag{3.10}$$

From their observation it appears that the retarding effect of the gypsum is due to its action as an electrolyte on a colloid solution, that is in this case, it prevents the hydrated aluminates from precipitating from its solution to form a gel. They found that with 2% of plaster of Paris ($CaSO_4 \cdot H_2O$) the hydrated substance could coagulate and separate as a solid less readily, with other concentrations of plaster, the separation of the solid occurs in a shorter time, and takes place more rapidly.

Effect of Fineness on Setting and Hardening

It is very important for the cement to be fine. In a cement of normal fineness, only about half the particles are hydrated when the cement has hardened. The reaction with water is always very incomplete because of the impervious coating formed about the grins; the larger the grain, of course, the greater will be the amount of the material anhydrate in the center. The anhydrate centers of the grains have functions that could not be fulfilled more cheaply by sand. Thus, it is evident that the finer cement allows the use of more sand in the mixture.

A thin mixture of cement and water salts is slower than one that contains less water because the particles are separated by the excess water. The speed of chemical change increases with the increase in temperature, and the setting and harding processes in Portland cement are no exceptions.

CHAPTER 4

The Chemical Analysis of Cement

Cement analysis is mainly done to control its quality which is to determine the composition and percentage of constituent of the compounds.

As described earlier, the principal compounds present in Portland cement are tricalcium silicates (C_3S), dicalcium silicate (C_2S), tricalcium aluminates (C_3A), and calcium alumina ferrite (C_4AF). Although the relative proportions of these compounds (particularly the first three) influence the properties of the cement, it is not possible to determine them by conventional chemical methods, nor is it yet possible to predict accurately the physical behavior of cement from a knowledge of its compound composition.

For some purpose, however, it is adequate that this is calculated from the analytical results expressed in terms of percentage of calcium oxide (CaO), silica (SiO_2), alumina (Al_2O_3), and iron oxide (Fe_2O_3).

At the expense of observant operatives and formulation of experimentally demonstrated principles by engineers and chemistry, there certain definite limitations have been established in compositions of cement. Within those limits, experience has shown that the mixture behaves satisfactorily in kilns and produces good cement; outside of those limits, it is also shown that trouble in burning may result in cement of inferior quality. This is true of any cement produced anywhere.

There are possible defects arising from unbalanced composition. For instance, if the lime content is too high, the extra lime does not come into a combination, and this may cause expansion and cracking of the mortar or concrete, which are made from the cement in question.

In addition to expressing the property or quality of cement in simply oxide forms, it can also be expressed in terms of the module which are the percentage ratios in weight of the oxides.

1. The hydraulic modulus (HM) is

$$\left(HM = \frac{\mathrm{CaO}}{\mathrm{SiO_2 + AL_2O_3 + Fe_2O_3}} \quad \%Wt \right). \tag{4.1}$$

   This expresses the hydraulic nature or property of cement. It should be between 1.8 and 2.2. If it increases beyond the limit when no more lime can be combined it will make the resultant cement unsoundness. If also decreased beyond certain limits, the cement can be called non-hydraulic.

2. The silica modulus (SM) is

$$\left(SM = \frac{\mathrm{SiO_2}}{\mathrm{AL_2O_3 + Fe_2O_3}} \quad \%Wt \right). \tag{4.2}$$

   The value should be in the range of 2.0–2.5. Cement with high silicate modulus hardens slowly, those with low silicate modulus set rapidly; that tends to have excellent early strength but low progression with increasing age. In the former case (with high SM) it may not give quite as high strengths as fresh, but tends to show a better progression with age.

Silica, alumina, and ferric oxide are likewise limited. If the lime content is fixed, and silica becomes too high, which may be accompanied by a decrease in alumina and ferric oxide, the temperature of burning will be raised and the special influence of the high lime is lost. If lime is too low which means an increase in the alumina and ferric oxide, the cement may become quick setting (reacts with water and hardens as fast as it is mixed) and contains large amounts of alumina compounds which appear to be of little value for their cementing qualities. Rapid setting is undesirable because the cement sets in so rapidly that it cannot be properly worked.

The magnesia (MgO) content should be limited, not to exceed 5%, because higher magnesia may be dangerous to the soundness (constancy in volume) of cement especially at the later ages for Portland cement, as shown in Table 4.1.

4.1 SPECIFICATION AND TESTS

Testing should be carried out according to relevant standard methods for the chemical test of cement.

Table 4.1: Approximate compositions of hydraulic cement with major components calculated as oxide

Cement	CaO	SiO_2	Al_2O_3	Fe_2O_3	SO_3	MgO
High-alumina	38	5	38	13	-	-
Hydraulic lime	60	20	8	2	-	1
Nature	45	25	5	4	2	10
Masonry	50	15	5	2	2	3
Oil well	63	21	5	6	2	2
Portland	64	21	6	3	3	2
Pozzolanic	45	30	12	4	2	1
Slag	50	26	12	2	2	1
Super sulphate	45	25	13	1	7	1

The tests applied directly to cement are generally concerned with fineness and with maximum or minimum limitations on certain oxides or calculated compounds consistent with the cement.

Indian standards installation specification for Portland cement:

$$\left(\frac{\mathrm{CaO}}{2.8\,\mathrm{SiO_2} + 1.2\,\mathrm{AL_2O_3} + 0.65\,\mathrm{FeO_3}}\right), \tag{4.3}$$

not greater than 1.02 and not less than 0.66. CaO, SiO_2, Al_2O_3, and Fe_2O_3 indicate their percentages in cement:

$$\left(\frac{\mathrm{Al_2O_3}}{\mathrm{FeO_3}}\right). \tag{4.4}$$

Other Requirements

1. Fineness. Residue on B.S 170 mesh test I sieve not to exceed 10%.

2. Setting time. Initial setting time, not less than 30 min, and final setting time not more than 12 h.

Since the Ethiopian standards specifications are not available in the institute's library or elsewhere in Bahr Dar, we could not present it here. We are of the opinion than the specifications are not far from those presented above.

For specifications and research purpose, the approximate compound compositions of Portland cement can be calculated by using the equation developed from the equilibrium data, as shown in Table 4.2.

Table 4.2: Percent oxide composition of Portland cement

Compound	Percentage
CaO	63.85
SiO_2	21.85
Al_2O_3	5.79
Fe_2O_3	2.86
MgO	2.47
SO_3	1.73

Thus, for *α* A/F% ratio of 0.64 or more, that is if $Al_2O_3/Fe_2O_3\%$ is of 0.64 or more.

$$\begin{aligned}
\%C_3S &= 4.07(\%C) - 7.60(\%S) - 6.7(\%A) - 1.43(\%F) - 2.85(\%SO_3) \\
\%C_2S &= 2.87(\%S) - 0.75(\%C_3S) \\
\%C_3A &= 2.65(\%A) - 1.69(\%F) \\
\%C_4AF &= 3.04(\%F).
\end{aligned}$$

When A/F% ratios are less than 0.64, a solid solution, of the aluminum and the iron phase (FSS), exists. Therefore,

$$\begin{aligned}
\%C_3S &= 4.07(\%C) - 7.60(\%S) - 4.48(\%A) - 2.86(\%F) - 2.85(\%SO_3) \\
\%C_2S &= 2.87(\%S) - 0.754(\%C_3S) \\
\%C_3A &= 3 \\
\%FSS &= 1.1(\%A) + 1.70(\%F) \\
\%C_4AF &= 0.
\end{aligned}$$

Cement chemistry employs a form of shorthand that has proven confusing to the non-informed. Chemical compounds are expressed as a sum of oxides using the first letters of the symbols of the non-oxygen components. Thus, tricalcium silicate, Ca_3SiO_5 or $3\,CaO \cdot SiO_2$ is written as C_3S. The symbols used include the following: $A = Al_2O_3$; $C = CaO$; $F = Fe_2O_3$; $H = H_2O$; $M = MgO$; $S = SiO_2$, sulphates are generally expressed in their usual form, $C_3A \cdot 3\,CaSO_4 \cdot 31\,H$ represents $3\,CaO \cdot AlO_3 \cdot 3\,CaSO_4 \cdot 31\,H_2O$.

4.2 METHODS OF ANALYSIS

An elemental analysis of cement is generally reported in terms of the oxide of the highest normal valence for each element. An exception is sulfide, which is reported as such metallic and ferrous iron determined only in special cases, reported as Fe and FeO, respectively. For example, an elemental analysis of different cements is shown in Table 4.3.

Table 4.3 indicates that cement can be made to have different properties by varying the composition or components. For example in Table 4.3, high-early strength (rapid hardening) gains strength faster than the ordinary Portland cement (regular Portland cement). Its initial and final setting times are the same as those of ordinary cement. It contains more tricalcium silicate and is more finely grounded. It is used for structures that are to be subjected to early use, e.g., rapier of bridges and roads, etc. It is more costly than ordinary cement.

The heat generated by cement while setting may cause the structure to crack in case of concrete. Therefore, as shown in Table 4.3, the heat generation is controlled by keeping the percentage of tricalcium aluminates (C_3S) and tricalcium silicate (C_3S) low. Its initial and final setting time is nearly the same as those of ordinary cement but the threat of its developing strength is very low.

The analysis of different kinds of cement also show that different compositions of cement may affect the properties of the cement. For example, in the analysis of high alumina cement; as shown in Table 4.4, there is a high percentage composition of alumina (Al_2O_3) and a lower percentage of calcium oxide (CaO) than the ordinary (regular) Portland cement. Advantages of using high alumina cement is its resisting property to the attack of chemicals, especially of sulphates and sea water. It also produces much higher ultimate strength than ordinary cement.

Analysis of natural pozzolana cement, which is shown in Table 4.4, indicates that it produces less heat of hydration and offers greater resistance to the attack of sulphate than ordinary Portland cement. It also reduces leaching of calcium hydroxide liberated during the setting and hydration process. The ultimate strength of this cement, but initial and final setting times, are the same. In these cases the natural pozzolana cement has greater percentage composition of SiO_2 and less percentage composition of CaO than the ordinary Portland cement, as shown in Table 4.5.

Table 4.3: Analysis of Portland cement (%)

	CaO	SiO_2	Al_2O_3	Fe_2O_3	MgO	Alkalioxide	SO_3
Regular Portland Cement							
Minimum	61.17	18.58	3.86	15.3	0.6	0.66	0.82
Maximum	66.92	23.26	7.44	6.18	5.24	2.9	0.26
Average	63.85	21.08	5.39	2.86	2.47	1.4	1.76
High-early strength: High C_3S							
Minimum	62.7	18.0	4.1	1.7	-	-	2.2
Maximum	67.5	22.9	7.5	4.2	-	-	2.7
Average	64.6	19.9	6.0	2.6	-	-	2.3
Low-heat of hardening lower $C_3S + C_3A$, higher C_2S_4 C_4AF							
Minimum	59.3	21.9	3.3	1.9	-	-	1.6
Maximum	61.5	26.4	5.4	5.7	-	-	1.9
Average	60.2	23.8	4.9	4.9	-	-	1.7

Table 4.4: Analysis of high-alumina cement (%)

Constituent	Percentage
SiO_2	3–11
Al_2O_3	33–44
CaO	35–44
Fe_2O_3	4–12
FeO	0–10

Table 4.5: Analysis of natural pozzolana cement (%)

Constituent	Percentage
SiO_2	30–64.71
Al_2O_3	10–33
Fe_2O_3	3.5–30
CaO	3–13
MgO	5
Na_2O	0.8–5
K_2O	0.8–9.5
TiO_2	0.5–1.35
SO_3	0.14–2.53

4.2.1 DETERMINATION OF COMPOUND COMPOSITION

Detailed analytical procedures are given below for each element of major importance in cement. For most of these elements, both a classical procedure, capable of the highest accuracy (e.g., gravimetric determination of SiO_2), and a rapid procedure, suitable for survey or control purposes (e.g., compound composition of hydraulic cement by x-ray diffraction), are given in Fig. 4.1.

Individual hydraulic cement can be identified by determination of their compound composition, through the use of an x-ray diffract meter or a light microscope. The former is used for the quantitative determination by measuring the intensity which is proportional to the electrometer deflection due to the reflected rays. But the latter is used for the qualitative determination of cement that is by looking at the crystal system, transmitted light, and incident light through the microscope. For example, the presence of C_3S (tricalcium silicate) and the crystal system observed is triclinic, and having the transmitted light a refractive index of $\alpha = 1.714$ and $\gamma = 1.717$, and an incident light with parallel grams, often containing spherical shapes (see Fig. 4.1 and Table 4.6).

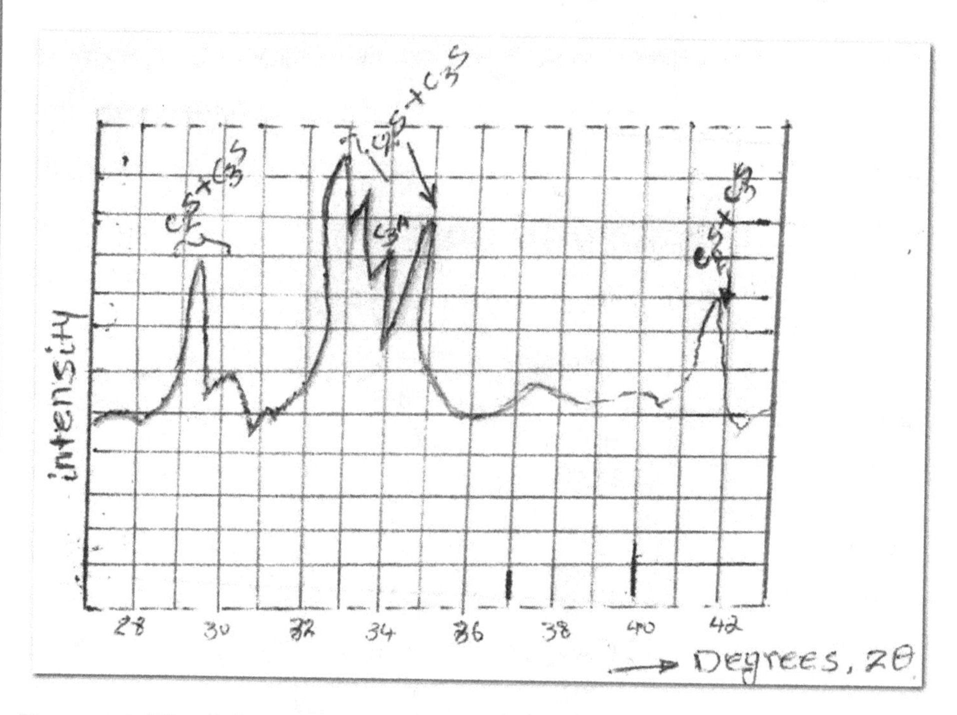

Figure 4.1: Hand-drawn x-ray diffraction pattern of a Portland cement (scanned copy by the author).

For examination by incident light, a suitable reagent that will differentiate present compounds is convenient. For example, dilute solutions of nitric acid which attack the silicate, potassium hydroxide, and aluminates are used. Magnesium which is the hardest substance is not attacked but it is easily identified because of its high release. The ferrite phase has the brightest reflectance and is readily identifiable. After compound identification has been obtained, reference can be made to the literature of suspected cements for composition information, as shown in Table 4.7, which indicates probable concentration range for major compounds present in selected cement.

A few compounds or groups of compounds can be determined by chemical extraction methods. These include free lime (calcium oxide plus calcium

Table 4.6: Microscopic data for major Portland cement compounds

Compound	Crystal System	Optical Characteristics Transmitted Light	Incident Light
C_3S	Triclinic	Negative indices $\alpha = 1.714$, $\gamma = 1.717$	Parallelogram often containing sperical shapes
β - C_2S	Monoclinic	Refractive indices $\alpha = 1.707$, $\beta = 1.715$, $\gamma = 1.730$	Irregularly shaped units tending toward sphericity
C_3A	Cubic	Refractive index n = 1.710	Irregular in shape or tending toward units having 45 or 90 corners
C_4AF	–	Refractive indeces $\alpha = 1.96$, $\beta = 2.01$, $\gamma = 2,04$	–
MgO	Cubic	Refractive index n = 1.736	Small angular units often displaying 45 and 90 corners
CaO	Cubic	Refractive index n = 1.83	Small render units

Table 4.7: Probable concentration range for major compounds present in selected cement

Type of Cement	Compoound Concentration Range for Cement								
	C_3S	C_2S	C_3A	C_4AF	CaO	Pozzolan	Slag	CA	C_2AS
High Alumina		10–15		0–20				60–65	2–20
Portland, Portland entertaining	10–60	20–30	0–15	0–15	0–2	0	0		
Portland blast furnace	8–45	15–45	9–'12	3–13	0–2	0	25–65		
Portland Pozzolan	8–50	16–50	0–12	0–2	0–2	15–35			
Slag				20–40	0	60–80			

hydroxide) and insoluble residue. There are methods available for free silica, magnesia alite, C_3S, and others. Water and carbon dioxide (carbonate) can be determined by the usual procedures. The sum of the latter two compounds is generally found by a loss on ignition determination, as given in the procedure below. Thus, loss, however, may be affected by other components present in the sample.

Procedures

Free lime:

- Prepare a methylene blue solution by dissolving 0.2 g of methylene blue solution in 10 mL of ethanol and adding it to 200 mL of isobutyl alcohol; filter.
- Prepare a thymol blue solution by dissolving 0.2 g of thymol blue in 10 mL of ethanol and adding it to 200 mL of isobutyl alcohol; filter.
- Introduce 100 mL of 3:20 ethyl acetoacetate-isobutyl alcohol mixture into each several 200-mL flasks. Introduce approximately 0.5 g of the sample or an amount of standard calcium oxide estimated to be equivalent to the free lime in the sample (about 0.005 g) into all the flasks, except the blank.
- Add 10 drops of freshly prepared 1% ethanolic sodium hydroxide solution. Reflux the standards, samples, and reagent blank for 30 min. Filter the solutions and wash each flask and residue with 50 mL of the hot isobutyl alcohol.
- Cool to room temperature and add 1 mL of thymol blue solution plus 8 drops of methylene blue solution to each flask. Titrate with 0.2 N perchloric acid in isobutyl alcohol to blue endpoint:

  $$\% \text{ free lime} = A(D - C) / E(B - C),$$

  where

  A = weight of standard calcium oxide used, in g

  B = volume of perchloric acid used to titrate A, in mL

  C = volume of perchloric acid used to titrate the blank, in mL

D = volume of perchloric acid used to titrate the sample, in mL

E = weight of sample, in g.

Insoluble residue:

- Transfer 1 g of the sample to a 250-mL covered beaker and add 25 mL of cold water. Also prepare a reagent blank. Swirl the mixture to disperse the cement in the water add 5 mL of concentrated HCl and swirl until the sample is dissolved. In some cases, it may be desirable to warm the solution slightly to ensure complete solution.

- Dilute the solution to approximately 50 mL with hot water. Transfer the beaker to the surface of a high temperature hot plate and bring the solution to incipient boiling.

- Digest the sample in the covered beaker for 15 min at a temperature just below boiling; this can be done by moving the beaker to the edge of the hot plate.

- Filter the hot solution using a glass rod into medium texture paper, catching the filter in a 400-mL beaker. Rinse the beaker with hot water and wash the contents of the filter paper thoroughly. Return the filter paper to the original beaker. Sulphur oxide can be determined on this filtrate.

- Add 100 mL of hot 1% aqueous NaOH solution and break up the filter paper with a glass stirring rod. Digest the sample at a temperature just below boiling for 15 min; stir the mixture several times during the digestion period.

- Add 2 drops of 0.2% ethanoic methyl; the red indicator changes from yellow to red. Add 5 drops HCl in excess. Filter on to medium texture paper.

- Scrub the beaker with a rubber policeman and rinse with hot water. Wash the residue on the filter paper 12–15 times with hot 2% aqueous NH_4NO_3 solution—this is to ensure the complete removal of all salt prior to the ignition of the sample.

- Transfer the paper and contents to a weighed platinum crucible. Dry and char the paper under an infrared lamp or over a fisher burner.
- Ignite in a muffle furnace at 950°C for 45 min. Cool in desiccators for 30 min and weigh.
- Calculate % insoluble residue to the nearest 0.01% by multiplying the weight, in g of the insoluble residue less the weight of the blank by 100.

Loss on ignition: Portland cement heats a number of crucibles equal to the number of samples to be analyzed at a temperature of 950 ± 50°C. Cool, weigh, and reheat to constant weight.

- Weight 1 g of each sample into individual crucible. Cover and ignite at 950 ± 50°C for 15 min. Transfer to a desiccator cooler, and weigh.
- Reheat the crucible for 5 min until it achieves a constant weight.
- Calculate % ignition is a loss to the nearest 0.1% multiplying the loss in weight, in g, by 100.

The loss of ignition is primarily due to water and carbon dioxide. However, oxidizable matter may also give rise to loss (organic matter) or gain (sulphide sulphur, etc.) in weight.

4.2.2 DETERMINATION OF MAJOR COMPONENTS

The chemical and physical properties of most elements have been reported to be almost entirely due to the major components of the cement. The analytical chemistry of cement has, thus, been concerned primarily with these elements present in major amounts. A recent investigation, however, indicates that the same element present in trace quantities in cement may have important effect out of proportion to their concentration.

Various instrumental methods can be employed for almost all of the elements present in cement, with varying degrees of accuracy. For example, x-ray spectrograph offers good accuracy for major components.

For major components in cement most instrumental techniques do not give analytical results on accuracy sufficient for referee acceptance purposes; thus, gravimetric and volumetric (i.e., wet chemical) methods are used for such analysis.

The procedures given below, applicable to the more common hydraulic cement, are capable of producing great accuracy for the cement or oxides described, provided that the effect of the more common interferences are recognized. Usual interferences caused by the presence of trace or rare element, or by application of this method to unusual products can only be detected by analyzing each precipitate spectro chemically.

Silicon dioxide: Silica can be best determined by gravimetric, colorimetric, or x-ray spectrographic methods. A method for the referee method is described in detail below.

Procedure

- Weigh 0.5000 g of the sample into a 100 mL platinum evaporating dish. Add 10 mL of cold water and whirl to disperse the solid in the water. While still swirling the sample, add 10 mL of concentrated HCl.
- Warm the platinum dish on steam bath until solution is complete. Any lump remaining should be broken up by using a rubber-tipped policeman.
- Evaporate the solution to dryness on a steam bath. Remove the dish from the steam bath and cool.
- Add 10 mL of concentrated HCl, cover the platinum dish, and allow it to digest for several minutes.
- Add an equal amount of water and return the covered dish to the steam bath for 10 min.
- Add an equal volume of hot water and filter through a medium texture paper. Receive the filtrate in a 400-mL beaker. Wash the restive on the filter paper ten times with hot 1.99 HCl and five times with hot water.
- Return the filtrate to the platinum evaporating dish and again evaporate to dryness.
- Bake the residue in an oven at 110°C for 1 h, cool, then add 15 mL of 1:1 HCl, and return the covered platinum dish to the steam bath for 10 min.
- Add an equal amount of hot water and filter through a medium texture paper into the 400-mL beaker. Repeat the washing of the precipitate, as

above. Reserve the filtrate for the determination of the ammonium hydroxide group.

- Place the filter paper containing the SiO_2 precipitate into a weighted platinum crucible. Dry under an infrared lamp until the filter paper is completely charred.
- Transfer the crucible without the lid to a clay triangle and ignite over a fish burner to remove all carbon. Cool; transfer the crucible (still without acid) to a muffle furnace.
- Remove the crucible to desiccators and cool for 30 min. Weigh and reheat to a constant weight.
- Moisten the ignited SiO_2 with a few drops of water. Add two or three drops of 1:1 H_2SO_4, 10 mL of 48% HCl and evaporate without boiling to near dryness. The small amount of H_2SO_4 remaining can be removed by holding the crucible with crucible tongs and continuously heating over a fisher burner.
- Finally, ignite to red heat over the fisher burner. Cool and transfer (without the lid) to a muffle furnace at 1100°C for 7 min.
- Weigh and then reheat to a constant weight. If the residue remaining after the treatment with HCl is greater than 0.002 g; reheat the SiO_2 procedure.
- The difference between this weight and the weight of the crucible plus the ignited SiO_2 previously obtained represents the amount of SiO_2 present in the sample. Calculate % SiO_2 to the nearest 0.01% by multiplying the weighting of SiO_2, less the weight of the blank, by 200.

Cement other than Portland may not be completely soluble in HCl and will give a high hydrofluoric acid residue. In such cases, begin the analysis again.

- Filter off the HCl insoluble material; wash, ignite, and fuse with quantity of $NaCO_3$ equivalent to about five times its weight.
- Return the melt to the beaker containing the acid-soluble material, dissolve using heat, and if necessary add more HCl.
- Evaporate the solution to dryness and dehydrate the silica as above.

Ammonium hydroxide group: The value of the NH_4OH group, R_2O_3, for element is usually assumed to include Al_2O_3 and for greater accuracy titanium oxide and phosphorus pentaoxide. Many other elements may be included, however, if present in the sample.

Procedure

- Add 0.5 g of fused potassium pyrosulphate to the residue from the silica determination in the UN covered crucible and warm just below red heat until the small amount of residue is dissolved in the melt.
- Cool and then place the crucible in the beaker containing the filtrate reserved from the silica determination.
- Heat the solution until the melt is dissolved.
- Remove and wash the crucible using a rubber policeman for this operation.

When analyzing Portland cement, if after repeating the silica determination two or three times a residue greater than 0.0020 g remains, it is probable that non-cementation material is present. In such a case, do not fuse the hydrofluoric acid residue found and proceed with determination using only the reserved filtrate.

- Dilute the solution to approximately 200 mL. Add 8–10 mL concentrated HCl and 3 ml of saturated bromine water over with a watch glass, and boil the solution to remove the bromine.
- The removal of bromine is complete when 1 drop of methyl red indicator permanently imparts a red color to the solution.
- While the solution is boiling, prepare a filter rock of funnels each felted with a medium texture paper.
- Use 600-mL beaker to receive the filtrates.
- Prepare a 2% NH_4NO_3 solution. Add 1 or 2 drops of 0.2% ethanoic methyl red indicator solution, and make distinctly alkaline with a few drops of 0.2% ethanoic methyl red indicator solution, and make distinctly alkaline with a few drops of 1:1 NH_4OH.
- Heat this NH_4NO_3 wash the solution in wash bottle.

- Remove the boiling bromine-free solution successively to a white-glazed tile; rinse and remove the watch glass, and add 1:1 NH_4OH until the first permanent precipitate begins to form.
- Add 2 drops of methyl red indicator.
- Add 1:1 NH_4OH dropwise, while stirring until the color changes from red to yellow.
- Add 2 drops in excess. Allow the precipitate to settle, and recheck the color of the supernatant liquid to make sure it is distinctly yellow.
- If an excess of NH_4OH has inadvertently been added, acidify with HCl, and again add NH_4OH to obtain the proper endpoint.
- Place the beaker on the hot plate; bring to a boil, for 30 s. Move the beaker to the side of the plate and allow the precipitate to settle. The temperature of the solution should be kept low enough to prevent any further boiling.
- While one precipitate is setting, complete the precipitation of any companion samples. Do not allow an appreciable time to elapse between precipitation and filtration.
- The solution must be distinctly yellow before filtration is started; if an orange color is evident add 1 drop 1:1 NH_4OH and stir. Filter the solution. Filtration is speeded if the precipitate on the bottom of the beaker is not disturbed until all the supernatant liquid has been transferred to the filter paper.
- Rinse the beaker once with hot NH_4NO_3. Wash the solutions, and proceed to the next samples.
- Wash the previously filtered precipitates without NH_4NO_3; wash solutions upon completion of each filtration. Each precipitate should be washed a total of four times. Cover and reserve the filtrate.
- Return each filter paper to the beaker in which the original precipitation took place. Prepare a filter rack with funnels containing medium texture paper.

- Add 10 mL of 1:1 HCl by pipette to each of the beakers. Immediately add 100 mL of boiling water to each beaker and stir the contents until the precipitate is dissolved.

- Re-precipitate the hydroxide from the near boiling solution as described above.

- Filter the solution, as above, expecting that at the conclusion of each filtration that the beaker will be scrubbed with a rubber policeman, and wash the inside of the beaker with NH_4NO_3 wash solution.

- After all filtrations are complete, wash each precipitate four times with hot NH_4NO_3 wash solution.

- Transfer the filter papers and precipitates to weighed platinum crucibles. Dry under an infrared lamp until the filter paper is completely charred. Transfer the crucible to a clay triangle and ignite over a fisher burner to remove all carbons.

- Ignite the uncover crucible in a muffle furnace at 1100°C for 30 min. Replace the covers and remove from the furnace.

- Cool in desiccators for 30 min; weigh and reheat to constant weight.

- Calculate % NH_4OH group in each sample to the nearest 0.01% by multiplying the weight of the precipitate, in g, less the weight of the blank by 200.

Aluminium oxide: Aluminium oxide is ordinarily calculated as the difference between % ammonium hydroxide group and % ferric oxide. The ammonium hydroxide group, however, generally includes several other oxides. For greater accuracy, or for comparison with direct instrumental results, the following calculation should be made:

$$\%\ Al_2O_3 = \%\ R_2O_3 - (\%\ Fe_2O_3 + \%\ TiO_2 + \%\ P_2O_5).$$

The amounts of manganese present do not lead to an error in this calculation for Portland cement.

Ferric oxide: Iron, in its oxidized forms, is usually determined by volume methods. Titration with potassium dichromate is the classical procedure for ferric oxide and is described below.

Procedure

- Prepare a saturated solution of mercuric chloride containing 1.229 g in 1878 mL of water.
- Dry a sample of potassium dichromate to constant weight at 200.
- After cooling in desiccators, weigh exactly 2.4570 g and transfer to 1-L volumetric flask.
- Dissolve in water and dilute to volume; 1 mL is equivalent to 0.004 g of Fe_2O_3.
- Dissolve 5 g of stannous chloride in 10 mL of concentrated hydrochloric acid, heat gently, and dilute to 100 mL with water. If the solution is not clean, boil with a few pieces of iron-free granulated tin until it clears. Transfer to a dropping bottle containing iron-free metallic tin.
- Weigh 1 g of sample into a 250-mL beaker. Add 40 mL of cold water, stir with a rubber policeman to disperse the sample, and add 10 mL of concentrated hydrochloric acid. Stir until the sample is decomposed.
- Rinse the policeman (stir on) with water, cover the beaker, and heat the sample to a boiling.
- Remove the beaker to a well-lit white surface; stir and add stannous chloride solution dropwise until the yellow color of ferric iron disappears and the solution becomes colorless. Then add 2 drops of the stannous chloride solution in excess.
- Rinse the stirrer and the inside walls of the beaker with water bath. Avoid prolonged standing at this time.
- Gently place Teflon-coated magnetic stirring bar into the beaker and place the beaker on a magnetic stirrer.
- Add 10 mL of saturated mercuric chloride solution and stir rapidly for 1 min. If little or no precipitation occurs, reduction of the iron was incomplete and the sample must be rejected.

- At the end of the 1 min period add 10 mL of 1:1 phosphoric acid and 2 drops of 0.3% aqueous barium diphenylamine sulphate indicator solution. Titrate to an endpoint at which 1 drop potassium dichromate solution causes an intense purple color to develop. If the color fades, add more potassium dichromate solution until a permanent purple color is obtained.
- Run a blank titration on the reagent, using the same procedure. If no purple color is obtained after adding 3 or 4 drops of the potassium dichromate solution, the blank should be assumed to be 0 mL:

  $\% \ Fe_2O_3 = E\ (V - U)\ 100$

  where

  $E = 0.004$

  V = volume of potassium dichromate solution need for a 1 g sample, in mL

  V = volume of potassium dichromate required for the blank, in mL.

Calcium oxide: Calcium, as determined by the usual gravimetric methods, generally includes strontium. A given weight of strontium oxide in the final ignited precipitate would be reported as the same weight of calcium oxide. It is believed that manganese contaminates the precipitate and leads to slightly high results.

Volumetric method such as complex metric titration with EDTA also determine strontium oxide as calcium oxide. However, due to the difference in the equivalent weights of calcium and strontium, 0.10% of SrO_3, for example, will be equivalent to about 0.05% of CaO. Among instrumental methods, it is generally agreed that x-ray spectrograph is at present the only way of accurately determining calcium in cement. In this instance strontium does not interfere.

The classical gravimetric procedure is given as follows.

Procedure

- Make the combined filtrate from the ammonium hydroxide group just acid, using 0.2% ethonlic methyl red indicator solution and 1:1 hydrochloric acid.

- Add 5 mL of 1:1 HCl in excess and evaporate the solution of 100 mL. The excess HCl is added to prevent formation of calcium sulphate precipitate; such precipitate is very difficult to get back into the solution.
- Place the cover beaker on a hot plate, and control the boiling. Place a small piece of filter paper inside the beaker. Hold it in place with a glass stirring rod.
- Heat the solution to boiling, remove from the hot plate, add 40 mL of saturated bromine water and then, immediately add 9–10 mL of 1:1 NH_4OH, and stir.
- Return to the hot plate and heat until vigorous boiling, again using a stirring rod and a piece of filter paper to prevent bumping.
- Place the beaker on a steam bath. Allow the precipitating hydrated manganese oxide to coagulate. A period of 60 min may be necessary to ensure the complete precipitation of a small amount of manganese dioxide.
- Filter the warm solution through a medium-textured paper. Rinse the beaker thoroughly into the funnel and wash the filter paper eight to ten times with hot water. Discard the precipitate and paper.
- Acidify the filtrate with hydrochloric acid and dilute to about 200 mL. The acidify can be checked with indicating pit paper.
- To remove bromine, heat the solution to boiling. Use the filter paper stirring rod technique described above in order to control the boiling. Check for the complete removal of bromine by occasionally adding a drop of methyl red indicator. When the solution remains permanently red after the addition of methyl red, the removal of the caroming can be considered complete.
- Remove the hot solution from the plate. Add 5 mL of concentrated HCl and, if necessary, 2 drops of methyl red indicator.
- Heat the solution almost to boiling, transfer to a beaker and add 25–30 mL of warm 5% aqueous ammonium oxalate solution.
- Add 1:1 NH_4OH dropwise, while stirring, until the color changes from red to yellow; a few drops in excess will do no harm, wash down the inside

wall of the filtration bench. Allow the solution to digest for 1 h, stirring occasionally during the first 30 min.

- Filter, receiving the filtrate in a 600-mL beaker. Wash down the inside of the beaker with cold 0.1% aqueous ammonium oxalate wash solution.
- Decant the wash liquid into the filter paper, retaining the rod in the beaker. Wash the filter paper three times with a cold wash solution.
- Invert the funnel and wash the precipitation from the paper into the original beaker, using a fine steam of water, be sure that none of the precipitation remains on the surface of the funnel.
- Add 10 mL of concentrated hydrochloric acid and 200 mL of hot water to the beaker that contains the precipitate. Stir until the solution of the precipitate is complete.
- Add two drops of methyl red indicator and 20 mL of hot 5% ammonium oxalate solution and transfer the beaker of nearly boiling solution to another beaker.
- Re-precipitate the calcium oxalate by neutralizing the acid solution with 1:1 NH_4OH, as outlined for the first precipitation. Digest for 2 h and then filter the solution through the original filter paper into the beaker containing the first filtrate.
- Wash the precipitate five times with a cold oxalate wash solution.
- Transfer the filter paper to a weighed platinum crucible. Dry and char the paper under an oven. Cool, then place the covered crucible in a slanting position oven a fisher burner and burn off the carbon.
- Heat the tightly covered crucible in a furnace at 1100°C for 1 h.
- Cool in a desiccator containing fresh magnesium perchlorate and a carbon dioxide adsorbent to make sure the calcium oxide does not pick up any moisture or carbon dioxide.
- Weigh after 30 min; then reheat to constant weight:

$$\% \text{ CaO} = X(Y - T)100/ZW,$$

where

X = molecular weight of CaO

Y = weight of the ignited CaO

Z = molecular weight of CaC_2O_2

W = weight of the original sample

T = weight of CaO found in blank.

Magnesium oxide: Magnesium can be determined by gravimetric and EDTA complexometric titration methods. The classical gravimetric procedure involving precipitations as the pyrophosphate is given as follows.

Procedure

- Make the filtrate from the CaO determination acid with HCl and evaporate to approximately 250 mL.
- Cool the beaker in a container of ice water to room temperature or below. Place the beaker on a magnetic stirrer and introduce Teflon-coated stirring bar into the beaker.
- Stir and add 10 mL of 25% aqueous diammonium hydrogen phosphate, $(NH_4)_2HPO_4$ solution.
- Add 30 mL of concentrated NH_4OH, very slowly at first, until heavy precipitation occurs, the remaining NH_4OH may be added more rapidly. The addition of NH_4OH is easily accomplished by using a 50-mL burnt and allowing the NH_4OH to flow out dropwise.
- Remove the Teflon stirring bar and rinse it with a small amount of 1:20 NH_4OH.
- Reheat with any companion samples. Allow the precipitating solutions to digest overnight.
- Filter through a retentive paper and rinse the inside of each beaker once with 1:20 NH_4OH.

- Wash the content of the paperback into the original beaker, using a fine steam of 1:20 NH_4OH. Use a minimum amount of this solution.
- To each beaker, add 10 mL of 1:1 HCl and 2 drops of 0.2% ethnic methyl red indicator solution, dilute with water to 100 mL, and add 1 mL of 25% NH_4HPO_4 solution. Transfer each beaker in turn to the magnetic stirrer, introduce the magnetic stirring bar, and add 25 mL of concentrated NH_4OH as outlined above.
- Remove the magnetic stirring bar, rinse, and allow each solution to stand for about 2 h.
- Filter the solution through the original filter paper.
- Wash the precipitate twice with 10-mL portions of aqueous NH_4NO_3 wash solution containing 100 g of NH_4NO_3 and 200 mL of concentrated NH_4OH per liter.
- Transfer the filter paper to weighed porcelain or platinum crucibles, char the paper under an infrared lamp, and remove the carbon by igniting over fisher burners.
- Ignite the precipitants in a muffle furnace at 1100°C for 1 h; cool in a desiccators and weigh.
- Reheat to a constant weight. The precipitate is magnesium pyrophosphate, $Mg_2P_2O_7$.

Calculate % MgO in each sample, to the nearest 0.01% using the following formula:

$$\% \text{ MgO} = (w - w)(0.620)(100)/0.5000,$$

where

w = weight of magnesium pyrophosphate, in g

w = weigh of magnesium pyrophosphate found in the blank, in g

0.362 = molecular ratio of 2 MgO : $Mg_2P_2O_7$

0.500 = weigh of sample analyzed, in g

For any accurate analysis it may be desirable to recover the calcium present in the $Mg_2P_2O_7$ precipitate.

Sulphur trioxide: Sulphur, as found in most types of cement, is present in the sulphate form. As such, it is determined gravimetrically below.

Procedure

- Transfer 1 g of the sample to a 250-mL beaker, add 20 mL of cold water, and disperse the sample by stirring with a rubber policeman.
- Add 5 mL concentrated HCl and stir until the sample is dissolved. In some cases it may be necessary to heat the solution slightly to achieve complete solution.
- Add 25 mL of water, using this water to rinse off the policeman. Rapidly heat the solution almost to a boiling, immediately remove to the surface of a steam bath, and digest for 15 min.
- Filter the hot solution through a fine texture paper, police the beaker, and wash the contents of the funnel thoroughly. Dilute the filtrate to 250 mL with water; add a small piece of filter paper, and a glass stirring rod to hold the paper in a place at the bottom of the beaker, and heat to boiling.
- Add drop by drop, from a pipette, 10 mL of hot 10% aqueous barium chloride solution. Continue the boiling until the precipitate is well formed.
- Return the solution to the bath and digest overtime. Make sure that the volume of the solution is maintained between 230 and 260 mL.
- Filter the hot solution through a retentive paper and wash the paper and contents ten times with hot water. Transfer the paper to a weighed porcelain crucible, dry and char under an infrared lamp or over a low flame, and burn off the carbon over a fisher burner. Ignite in a muffle furnace at 800–900°C for 45 min, cool in a desiccator, and weigh as $BaSO_4$.
- Calculate % sulphur trioxide as follows:

  % sulphur trioxide = (w – w) (0.343) (100)/ws,

  where

w = weight of $BaSO_4$, in g

w = weight of $BaSO_4$ found in blank, in g

0.343 = mole ratio of SO_3: $BaSO_4$

ws = sample weight.

Bibliography

[1] *World Book Encyclopedia*, Field Enterprises Educational Corporation, 1961.

[2] *The McGraw-Hill Encyclopedia of Science and Technology*, McGraw-Hill, 1971.

[3] Robert B. Leighou, Chemistry of engineering materials, *Journal of Chemical Education*, Tokyo, 1953.

[4] Irving Steist, *Hand Book of Adhesive*, Springer, 1965. DOI: 10.1007/978-1-4613-0671-9.

[5] S. N. Ghosh, Ed., *Advances in Cement Technology*, Elsevier, India, 1983. DOI: 10.1016/C2013-0-06042-2.

[6] H. S. Stuuman, *Science and Intention and Invention*, New York, 1983.

[7] Surendra Singh, *Engineering Materials*, Stosius Inc., 1979.

[8] S. N. Ghosh, Ed., *Advances in Cement Technology*, Elsevier, India, 1981.

[9] S. D. Shukla and G. N. Pandey, *Textbook of Chemical Technology*, Vikas Publishing House Private, India, 1978.

[10] Don A. Watson, *Construction Materials and Processes*, McGraw-Hill, 1978.

[11] G. Barnbrook, *Concrete Practice*, Cement and Concrete Association, 1975.

[12] M. M. Uppal and S.C. Bhatia, *Engineering Chemistry*, Khanna Publishers, New Delhi, 1983.

[13] Roger C. Griffin, Ed., *Technical Methods of Analysis*, McGraw-Hill, 1927.

[14] F. D. Snell and L. S. Ettre, *Encyclopedia of Industrial Chemistry Analysis*, Interscience Publishers, 1974.

[15] R. Norris Shreve, *Chemical Process Industries*, 3rd ed., McGraw-Hill, 1967.

[16] Hoang-Anh Nguyen et al., Sulfate resistance of low energy SFC no-cement mortar, *Construction and Building Materials*, 102:239–243, 2016. DOI: 10.1016/j.conbuildmat.2015.10.107.

Author's Biography

TADELE ASSEFA ARAGAW

Tadele Assefa Aragaw, born in 1989, is a member of the Faculty of Chemical and Food Engineering, Department of Environmental Engineering. He obtained his MSc. in 2017 in Chemical Engineering (Environmental Engineering) at Bahir Dar Institute of Technology, Bahir Dar University, Bahir Dar, Ethiopia, and conducted research at Bahir Dar Institute of Technology, Faculty of Chemical and Food Engineering. His first research interest is the use of phytoplankton (microalgae) to investigate the role of microalgae in the purification of wastewater and algae biomass production for biofuel. His second research interest is the use of geological clay materials, specifically, kaolin, for industrial application and synthesis of adsorbent as low-cost for textile dye wastewater. Tadele Assefa Aragaw has published eight Scopus indexed articles, book series, and proceedings in the research areas previously described.

CPSIA information can be obtained
at www.ICGtesting.com
Printed in the USA
FSHW020754140220
67021FS

9 781681 736532